Software Test Plans

A How To Guide for Project Staff

David Tuffley

To my beloved Nation of Four
Concordia Domi – Foris Pax

And oftentimes excusing of a fault doth make the fault the worse by the excuse. William Shakespeare.

Acknowledgements

I am indebted to the Institute of Electrical and Electronics Engineers on whose work I base this book, specifically IEEE Std 829.

I also acknowledge the *Turrbal* and *Jagera* indigenous peoples, on whose ancestral land I write this book.

Contents

- A. Introduction 5
  - A.1. Scope 5
  - A.2. References 7
  - A.3. Definitions and acronyms 7
  - A.4. Responsibilities 10
- B. Standard / procedure description 11
  - B.1. Test plan 11
    - B.1.1. Purpose 11
      - B.1.1.1. Software life cycle context 11
      - B.1.1.2. Use 12
    - B.1.2. Format & content 14
      - B.1.2.1. Cover page 14
      - B.1.2.2. Page marking 15
      - B.1.2.3. Test Plan Table of Contents 15
      - B.1.2.4. Identification and Purpose (section 1) 16
      - B.1.2.5. Introduction (section 2) 17
      - B.1.2.6. Test Items (section 3) 18
      - B.1.2.7. Features to be tested (section 4) 19
      - B.1.2.8. Features not to be tested (section 5) 19
      - B.1.2.9. Approach (section 6) 19
      - B.1.2.10. Item pass/fail criteria (section 7) 21
      - B.1.2.11. Suspension criteria & resumption requirements (section 8) 21
      - B.1.2.12. Test deliverables (section 9) 21
      - B.1.2.13. Testing tasks (section 10) 22
      - B.1.2.14. Environmental needs (section 11) 23
      - B.1.2.15. Responsibilities (section 12) 23
      - B.1.2.16. Staffing & training needs (section 13) 24
      - B.1.2.17. Schedule (section 14) 24
      - B.1.2.18. Risks & contingencies (section 15) 25
      - B.1.2.19. Approvals (section 16) 26

Contents

B.2. Test Design specification 27
   B.2.1.   Purpose 27
   B.2.2.   Software life cycle context 27
   B.2.3.   Use 27
   B.2.4.   Format & content 28
      B.2.4.1. Cover page 28
      B.2.4.2. Page marking 28
      B.2.4.3. Identification and Purpose (section 1) 29
      B.2.4.4. Features to be tested (section 2) 29
      B.2.4.5. Testing approach (section 3) 29
      B.2.4.6. Test identification (section 4) 30
      B.2.4.7. Feature pass/fail criteria (section 5) 30

B.3. Test case 31
   B.3.1.   Purpose 31
   B.3.2.   Software life cycle context 31
   B.3.3.   Use 31
   B.3.4.   Format & content 32
      B.3.4.1. Cover page 32
      B.3.4.2. Page marking 32
      B.3.4.3. Identification and Purpose (section 1) 33
      B.3.4.4. Test items (section 2) 33
      B.3.4.5. Input specifications (section 3) 33
      B.3.4.6. Output specifications (section 4) 34
      B.3.4.7. Environmental needs (section 5) 34
      B.3.4.8. Special procedural requirements/rules (section 6) 35
      B.3.4.9. Intercase dependencies (section 7) 35

B.4. Test Procedure specification 36
   B.4.1.   Purpose 36
   B.4.2.   Software life cycle context 36
   B.4.3.   Use 36
   B.4.4.   Format & content 37
      B.4.4.1. Cover page 37
      B.4.4.2. Page marking 37
      B.4.4.3. Identification and Purpose (section 1) 38
      B.4.4.4. Purpose (section 2) 38
      B.4.4.5. Special requirements (section 3) 38
      B.4.4.6. Procedure steps (section 4) 38

B.5. Test Item Transmittal report 40
   B.5.1.   Purpose 40

Contents

B.5.2. Software life cycle context 40
B.5.3. Use 40
B.5.4. Format & content 40
    B.5.4.1. Cover page 40
    B.5.4.2. Page marking 41
    B.5.4.3. Identification (section 1) 41
    B.5.4.4. Transmitted Items (section 2) 41
    B.5.4.5. Location (section 3) 42
    B.5.4.6. Status (section 4) 42
    B.5.4.7. Approvals (section 5) 42

B.6.   Test log 43
B.6.1. Purpose 43
B.6.2. Software life cycle context 43
B.6.3. Use 43
B.6.4. Format & content 44
    B.6.4.1. Cover page 44
    B.6.4.2. Page marking 44
    B.6.4.3. Identification (section 1) 45
    B.6.4.4. Description (section 2) 45
    B.6.4.5. Activity & event entries (section 3) 45

B.7.   Test incident report 47
B.7.1. Purpose 47
B.7.2. Software life cycle context 47
B.7.3. Use 47
B.7.4. Format & content 48
    B.7.4.1. Cover page 48
    B.7.4.2. Page marking 48
    B.7.4.3. Identification (section 1) 49
    B.7.4.4. Summary (section 2) 49
    B.7.4.5. Incident description (section 3) 49
    B.7.4.6. Impact (section 4) 50

B.8.   Test Summary report 51
B.8.1. Purpose 51
B.8.2. Software life cycle context 51
B.8.3. Use 51
B.8.4. Format & content 52
    B.8.4.1. Cover page 52
    B.8.4.2. Page marking 52
    B.8.4.3. Identification (section 1) 53

Contents

B.8.4.4.	Summary (section 2)	53
B.8.4.5.	Variances (section 3)	53
B.8.4.6.	Comprehensive assessment (section 4)	53
B.8.4.7.	Summary of results (section 5)	54
B.8.4.8.	Evaluation (section 6)	54
B.8.4.9.	Summary of activities (section 7)	54
B.8.4.10.	Approvals (section 8)	55

A. Introduction

This standard defines testing documentation. It is based on IEEE Std 829 - Standard for Software Test Documentation.

A.1. Scope

This standard describes the complete document set required for the conduct of the test function. It describes:

- the software development context in which each document shall be created and
- minimum requirements for format and content.

The following describes the test documents

- **Test Plan:** a document describing the scope, approach, resources and schedule of testing activities.
- **Test Design Specification**: a document that provides details of the test approach in terms of the features to be covered, the test cases and procedures to be used and the pass/fail criteria that will apply to each test. The test design specification forms the entry criteria for the development of Test Procedures and the specification of Test Cases on which they operate.

Software Test Plans

- **Test Case**: a document specifying actual input values and expected outputs. Test cases are created as separate documents to allow their reference by more than one test design specification and their use by many Test Procedures.

- **Test Procedure**: a document describing the steps required to prepare for, run, suspend and terminate tests specified in the test design specification. As an integral part of the test the document specifies the test cases to be used. Test procedures are created as separate documents as they are intended to provide a step by step guide to the tester and not be cluttered with extraneous detail.

- **Test Item Transmittal Report**: a document identifying the test items being transmitted for testing.

- **Test Records**: a suite of documents which record the results of testing for the purposes of corrective action and management review of the effectiveness of testing. Test records are represented as:

- **Test Log**: a document used by the test team to record what happened during testing. The log is used to verify that testing actually took place and record the outcome of each test (i.e. pass/fail).

- **Test Incident Report**: a report used to document any event that occurs during testing that requires further investigation. The creation of a Test Incident Report triggers corrective action on faults by the development team at the completion of testing.

- **Test Summary Report**: a management report summarising the results of tests specified in one or more test design specifications. This document informs management of the status of the product under test giving an indication of the quality of software produced by the development team.

A.2. References

[1] IEEE Std 829 - Standard for Software Test Documentation.

A.3. Definitions and acronyms

Acceptance Testing: the process of evaluating the finished product to verify compliance with the customer requirements specification. Acceptance testing is carried out by the customer using real data in the operational environment.

Black Box Testing: the input/output behaviour of a module is considered only. The test design specification is derived from the module functional specification.

Advantage: simplifies testing.

Drawback: testing cannot be complete unless all possible combinations of inputs are tested.

Bottom Up Testing: a testing strategy that starts with modules at the lowest level (modules that do not call other modules) and simulates superordinate modules with drivers.

Software Test Plans

Advantages: tests can be realistic - driver modules can easily simulate test data.

Entry Criteria: the conditions that must exist before a task can commence. For example: prior tasks complete, deliverables complete, resources available and work authorization in place.

Integration Testing: the process of testing groups of modules to evaluate their correct interworking.

Software Feature: a characteristic of a software product, the proper functioning of which can be verified by a test strategy (for example, transaction rate, software function).

System Test: the process of stress testing a software product in the target environment prior to delivery to a customer.

Test Auditability: an attribute of a testing process that provides objective evidence (i.e. formal records) to support independent verification that the test was carried out in compliance with the test design specification.

Test Case: documentation describing inputs, expected results and execution conditions for a test item.

Test Design Specification: a document describing the sequence of actions for the execution of a test.

Test Incident Report: a document describing any event that occurs during testing that requires further investigation.

Test Item: a software and/or hardware component that is tested as a unit.

Software Test Plans

Test Log: a chronological record of events that occur in the execution of a test.

Test Procedure Specification: a document describing the sequence of actions required for the execution of a test and the test data sets to be employed.

Test Summary Report: a document summarising testing activities and results and providing evaluations of corresponding test items.

Thread Testing: a system is segmented into demonstrable functions - threads. Each module in a thread is coded and tested such that the complete thread can be verified.

Advantage: Critical functions can be tested early.

Top Down Testing: a testing strategy that starts at the top module and works down. Stub modules are used to simulate incomplete subordinate software modules.

- Advantage: progressive testing - eliminates the need for integration testing. Also provides for early validation of the design.

- Drawback: tests may not accurately verify performance - stub modules may not effectively simulate data flow.

White Box Testing: the structure of the program is examined and the test data is derived from the program's logic such that every statement in the program is executed at least once, a subset of every path in the program is transitioned at least once and each decision is made at least once.

Advantage: provides for more thorough testing.

Drawback: in a complex program it is impossible to transition all paths.

A.4. Responsibilities

These following staff have responsibilities associated with this standard.

Project Manager:

- Test Strategy and the planning of testing activity on a project.
- Ensuring test software and documentation is under configuration management control.

Manager QA:

- Ensuring that all projects comply with this Standard.

B. Standard / procedure description

B.1. Test plan

B.1.1. Purpose

The purpose of the test plan is to describe:

- the scope of testing in terms of the software features to be tested
- the approach to testing in terms of testing methods, tasks and deliverables
- testing logistics in terms of schedules, resource requirements and individual responsibilities.

B.1.1.1. Software life cycle context

The primary objective of the Test Plan is to provide a strategy to verify that the software product functionality described in the Software Requirements Specification has been delivered. The Test Plan is therefore developed upon completion of the Software Requirement Specification.

The Test Plan is succeeded by test design specifications, test cases and test procedures which provide detail of the conduct of individual tests.

Software Test Plans

B.1.1.2. Use

This section describes the purpose of the test plan and its application to various types of testing.

The purpose of the Test Plan is to:

- Demonstrate to management and the quality assurance function that the product is to be adequately tested.
- Document the funding and general resource requirements for the purposes of management approval.
- Provide a baseline for test group performance measurement.
- Provide criteria against which the customer will formally accept the product and
- Support regression testing of the product.

The Test Plan may be prepared to guide the conduct of acceptance testing, system testing or integration testing. Test plans are not required for unit testing which shall be performed by the development team. Unit tests are conducted in accordance with strategies described by the designer in the Software Design Description.

B.1.1.2.1. Acceptance test planning

The Acceptance Test Plan shall be prepared by the customer. It describes how the finished product will be tested for compliance with the customer's requirements specification. Acceptance testing is carried out by the customer using real

Software Test Plans

data in the operational environment. That is, the actual software, hardware, personnel and procedures that the system was designed to work with are used. Acceptance tests normally fall into the class of functional tests the goal of which is to verify that all functions specified in the System Requirements Specification are operational. The entire system is treated as a black box. No knowledge of the internal structure of the system is used in the design of test cases. The test designer works from a user manual or software/system requirements specification and focuses on the input/output behaviour of each function described.

B.1.1.2.2.System/Integration test planning

The System/Integration Test Plan shall be prepared by the project test group. The plan describes the process of integrating component modules and progressively testing module interaction. This process is concerned with:

- finding errors which normally result from unanticipated interactions between system modules
- validating that the overall system provides the functions specified by the client.
- validating that the product delivers its pre-specified performance (e.g. transactions per second). The objective of performance or stress testing is to make the product fail and, in so doing, ensure that failure occurs outside specified limits and that failure modes are acceptable to the client (e.g. do high transaction rates

cause total system failure or merely reduce response time?).

The System/Integration Test is prepared with a knowledge of the internal structure of the software (e.g. the product structure chart is known). This means that the test designer must work from Software Architecture Specifications and Interface Control Documents. In this context individual modules are viewed as black boxes and their interconnection and rules of interface are viewed as a white box.

B.1.2. Format & content

This section provides the minimum requirements for format and content of a Test Plan. All sections described here must be provided in Test Plan documents. Issues relevant to particular test strategies that are not provided by this outline may be included as required by the Tester and approved by the Project Manager. Figure 1 provides the standard Test Plan Document Outline.

B.1.2.1. Cover page

The cover page shall include:

- Project Title.
- Document Number.
- Volume Number.
- Document Class: `Test Plan'

Software Test Plans

- System Name.
- Revision Number.
- Author.
- Approval Authority.
- Approval Signature.
- Approval Date.

B.1.2.2. Page marking

Each page is to be marked with:

- Document Class: `Test Plan'
- System Name
- Document Number
- Revision Number
- Page Number

B.1.2.3. Test Plan Table of Contents

```
1.  Identification and Purpose
    1.1   Identification of Client and project
    1.2   Purpose
2.  Introduction
    2.1   Objectives
    2.2   Scope
    2.3   References
    2.4   Document overview
3.  Test items
4.  Features to be tested
```

Software Test Plans

```
5.  Features not to be tested
6.  Approach
    4.1  Tools and techniques
    4.2  Test activities
    4.3  Test coverage
    4.4  Constraints
7.  Items pass/fail criteria
8.  Suspension and redemption
9.  Deliverables
10  Tasks
11  Environmental needs
12  Responsibilities
13  Staffing and training needs
14  Cost schedule
15  Risks and contingencies
16. Approvals
```

Table 1. Test Plan Document Outline.

B.1.2.4. Identification and Purpose (section 1)

B.1.2.4.1.Identification of the Client and Project (section 1.1)

Formally describe the client and project.

B.1.2.4.2.Purpose (section 1.2)

Describe the purpose of the document, for example:

The purpose of this document is to:

- describe the activities required to prepare and conduct the XYZ system test,
- describe the approach to testing in terms of testing methods, tasks and deliverables,
- describe testing logistics in terms of schedules, resource requirements and individual responsibilities.

B.1.2.5. Introduction (section 2)

B.1.2.5.1. Objectives (section 2.1)

Define the overall objectives of the test activity.

Example:

- ..to determine that system XYZ satisfies all requirements specified in system requirement specification ABC.
- ..to identify the performance limitations of system XYZ under abnormal operating conditions.
- ..to verify that software product XYZ is a correct and complete implementation of software design description ABC.

B.1.2.5.2. Scope (section 2.2)

Summarise the scope of testing as described by this document.

B.1.2.5.3. References (section 2.3)

List the reference documents used in the preparation of the plan (for example, development plan, quality plan, configuration management plan, other test plans, this standard).

B.1.2.5.4. Document overview (section 2.4)

Describe the organisation and content of the remainder of the document.

B.1.2.6. Test Items (section 3)

Provide the following test item information:

- **Identification**: test item name and revision level (e.g. hardware model number, program/module name and revision level).
- **Media**: the transmittal media on which the item and/or its design description is held (e.g. tape, floppy disk).
- **Documentation**: identify all documentation that describes the test item. In particular identify documents that provide test criteria. For example:
- System/Software Requirements Specification
- Software Design Description
- User Guide
- Systems Operations Manual

- Installation Manual
- **Open Test Incident Reports**: identify any unresolved test incident reports raised against the test item as a result of previous testing. This paragraph is required to ensure that all faults are rectified.
- **Excluded Test Items**: identify items that are to be specifically excluded from testing.

B.1.2.7. Features to be tested (section 4)

- **Functionality**: describe the major areas of functionality to be tested. Note that the scope of the feature described is usually determined by that of the test design specification which verifies its proper functioning. For example, Network Configuration Data Management, Critical Path Scheduling Utility.
- **Test Design Specification ID**: identify the test design specifications that have been or will be developed to describe the detailed test sequences and data.

B.1.2.8. Features not to be tested (section 5)

Identify functions that are not to be tested and specify reasons.

B.1.2.9. Approach (section 6)

This section describes the overall approach to testing. Further details of testing tasks and equipment are provided in

Software Test Plans

succeeding sections. For each major group of features or feature combinations provide:

B.1.2.9.1. Tools & techniques (section 6.n.1)

- **Technique**: describe testing technique to be employed. For example, white box, black box, top down, bottom up and thread testing.
- **Test Bed:** describe test equipment and software required to adequately test the function.

B.1.2.9.2. Test activities (section 6.n.2)

- **Tasks**: provide overview statements of work describing major testing tasks to be performed.

B.1.2.9.3. Test coverage (section 6.n.3)

- **Comprehensiveness**: the degree of test comprehensiveness expressed in terms of (for example):
- the percentage of total statements that will be executed at least once.
- the exception conditions that will be exercised including null, zero and empty conditions.
- **Completion Criteria:** the criteria that will signify the end of testing (for example error frequency).

- **Test Criteria:** identify the source documents from which the Test design specifications will be developed.

B.1.2.9.4. Constraints (section 6.n.4)

Identify constraints on testing such as test item availability, testing resource availability and deadlines.

B.1.2.10. Item pass/fail criteria (section 7)

Specify the criteria to be used to determine whether each test item has passed or failed testing.

B.1.2.11. Suspension criteria & resumption requirements (section 8)

- **Suspension:** specify the criteria to be used to suspend all or a portion of the testing activity on a test item.
- **Resumption:** specify the testing activities that must be repeated when testing is resumed (for example, re-initialization of intermediate data structures). Note that if testing is suspended it may not be automatically assumed that all previous tests have passed.

B.1.2.12. Test deliverables (section 9)

- **Documentation:** identify the documentation that will control the conduct of testing and the recording of results with particular focus on test auditability. For example:

Software Test Plans

- Test design specifications
- Test procedure specifications
- Test case specifications
- Test logs
- Test incident reports
- Test summary reports
- **Tools**: identify tools that are to be developed to facilitate testing. For example:
- Drivers
- Stubs
- Test Equipment
- **Data:** identify data that are to be developed to facilitate testing. For example:
- Test input data
- Test output data

B.1.2.13. Testing tasks (section 10)

Describe the set of tasks necessary to prepare for and perform testing. Provide:

- Statements of work.
- Special skills required.

B.1.2.14. Environmental needs (section 11)

Describe the required properties of the test environment including:

- Hardware configuration
- Software configuration
- Communications system configuration
- Physical and logical security
- Requirements for consumable supplies
- Office facilities
- Air conditioning.

B.1.2.15. Responsibilities (section 12)

Identify the organizational groups responsible for:

- Managing testing activities and reporting on test results.
- Designing tests and delivering test design specifications.
- Providing and preparing the test environment.
- Executing tests and recording test results.
- Witnessing tests and certifying test results.
- Undertaking corrective action on Test Incident Reports.
- Monitoring corrective action to ensure that all open Test Incident Reports are followed up.
- Maintaining test results for the purposes of overall process improvement (e.g. statistical process control).

Software Test Plans

Note that this task is typically the responsibility of the quality assurance function.

B.1.2.16. Staffing & training needs (section 13)

Identify staffing and training needs in terms of:
- Job descriptions
- Skills required
- Skills available
- Levels of training required to meet specified skill levels
- Training options (e.g. in-house, external, requirements for development of specialized training)

B.1.2.17. Schedule (section 14)

Specify testing schedules and costs in terms of:
- Work breakdown structure.
- Critical path network providing:
- Tasks.
- Milestones (e.g. receipt of test items, completion of test documentation, test completion, certification of test results, transitioning of test items into the production environment)
- Elapsed time for each task.
- Task dependencies.

- Resource utilization (people and equipment) - how much for how long.

Note that costs may be expressed in terms of working days or dollars. Cost estimates should represent a further refinement of those presented in the project plan.

B.1.2.18. Risks & contingencies (section 15)

Risk is the potential for realization of unwanted, negative consequences of an event (William D. Rowe, An Anatomy of Risk).

A future event or outcome is considered a risk if:

- a loss of time, money, product quality or product functionality is associated with it
- uncertainty or chance is involved
- some choice is involved. That is, action can be taken now to avoid a risky event or reduce the magnitude of the associated loss.

This section identifies risk and provides risk management strategies. For example:
- Critical test items
- Risk assessment expressed as:
- risk scenarios
- risk probability
- magnitude of loss
- schedule impact (critical path assessment)

- Risk management plan - choices that can be made now to avoid risk.

- Budget allocation for risk (e.g. provision for additional testing resources). For example:

Critical Test Item - XYZ Subsystem
Risk Scenario
In view of the previous poor delivery performance of ABC Co. a 90% probability exists that ABC's IJK Subsystem will not be delivered on time thus delaying IJK/XYZ integration testing.
Potential Loss
Should this outcome be realised a loss of $3000 per day will be incurred through idleness of the test team.
Management Plan
A IJK test simulator will be built by the test team to guarantee that the XYZ Subsystem testing proceeds on schedule.
Budget Allocation
The estimated cost of the IJK Simulator is $15,000. This cost will be recovered in the prevention of a five day delay.

B.1.2.19. Approvals (section 16)

Identify all the names and position of the people who are to approve the test plan and reserve a space for signatures.

B.2. Test Design specification

B.2.1. Purpose

To further refine the description of the process to be used to test a feature or group of related features.

B.2.2. Software life cycle context

The test design specification flows from the test plan as the software design description flows from the software requirements specification. The test plan describes what must be tested. The test design specification describes how it is tested. The entry criteria for test design specification preparation are therefore the completion of the software architecture specification and interface control documents.

B.2.3. Use

This document is used as a basis for the specification of test procedures and test cases. An analogy can be drawn to the relationship between a software design description and a software program. The test design specification describes the structure and logic of the test procedure/test case and the test procedure/test case documents describe its implementation.

B.2.4. Format & content

B.2.4.1. Cover page

The cover page shall include:
- Project Title
- Document Number
- Volume Number
- Document Class: `Test Design Specification'
- System Name
- Revision Number
- Author
- Approval Authority
- Approval Signature
- Approval Date
- Reference to Associated Test Plan

B.2.4.2. Page marking

Each page is to be marked with:
- Document Class: `Test Design Specification'
- System Name
- Document Number
- Revision Number

- Page Number

B.2.4.3. *Identification and Purpose (section 1)*

As per the Test Plan document

B.2.4.4. *Features to be tested (section 2)*

- Identify the test items covered by the specification.
- Identify the feature or feature combinations to be tested.
- For each feature provide a reference to the requirement in the system/software requirement specification or software design description.

B.2.4.5. *Testing approach (section 3)*

Provide refinements to the broad approach described in the test plan. Specify:

- Specific test techniques to be used.
- The method of analyzing test results (e.g. automated or manual methods of comparing actual with expected test results).
- The reason for selection of various test cases.
- Summary of common attributes and environmental needs of test cases.

B.2.4.6. Test identification (section 4)

For each feature to be tested provide:

- Test procedure identifiers and descriptions.
- Test case identifiers and descriptions.
- The pass/fail criteria.

B.2.4.7. Feature pass/fail criteria (section 5)

For each feature specify:

- The criteria for passing and or failing

B.3. Test case

B.3.1. Purpose

To describe the expected output behaviour of a test item given input data sets, data rates and environmental conditions.

B.3.2. Software life cycle context

The test case flows from the test design specification. The test design specification describes the generic types of data to be applied in the test. The test case describes values, value ranges, tolerances and data entities to be applied. The entry criteria for test case preparation is therefore the completion of the test design specification, system/software requirement specification, software architecture specification, interface control document and software design description.

B.3.3. Use

In conjunction with the test procedure specifications, test case data is used by the tester to verify the expected input/output behaviour of a test item. Test cases are described as separate entities to allow for reuse by many test procedure specifications.

B.3.4. Format & content

B.3.4.1. Cover page

The cover page shall include:

- Project Title
- Document Number
- Volume Number
- Document Class: `Test Case Specification'
- System Name
- Revision Number
- Author
- Approval Authority
- Approval Signature
- Approval Date

B.3.4.2. Page marking

Each page is to be marked with:

- Document Class: `Test Case Specification'
- System Name

Software Test Plans

- Document Number
- Revision Number
- Page Number

B.3.4.3. Identification and Purpose (section 1)

As per the Test Plan document

B.3.4.4. Test items (section 2)

Identify the test items and associated features to be exercised by the test case.

Where appropriate, provide references to the following test item documentation:

- System/Software Requirements Specification
- System Architecture Specification
- Interface Control Document
- Software Design Description
- User Manual
- Installation Guide

B.3.4.5. Input specifications (section 3)

Specify each input required to execute the test case. Describe data in terms of:

- Data discrete values and ranges.

Software Test Plans

- Tolerances.
- Transaction file names.
- Terminal messages.
- Values passed by operating systems.
- Timing relationships between inputs.

B.3.4.6. *Output specifications (section 4)*

For each set of inputs describe the expected outputs in terms of:

- Exact values.
- Tolerances (if applicable).
- Response times.

B.3.4.7. *Environmental needs (section 5)*

Describe the requirements of the hardware/software environment for correct execution of the test case. Provide:

- Hardware configurations.
- System and application software configurations including compilers, operating systems, simulators and test tools.
- Specific skills required of test staff.

B.3.4.8. *Special procedural requirements/rules (section 6)*

Provide any special constraints on the test procedures that execute this test case. For example:

- Set up.
- Operator intervention.
- Measurement of output.

B.3.4.9. *Intercase dependencies (section 7)*

If applicable list the test cases that must be executed prior to this test case and describe the nature of dependencies.

B.4. Test Procedure specification

B.4.1. Purpose

To specify the set of steps required to evaluate the features of a test item. This involves specifying how a set of test cases will be applied.

B.4.2. Software life cycle context

The test procedure flows from the test design specification. The test design specification describes the logical flow of test activities. The Test Procedure describes the implementation in terms of step sequences. The entry criteria for test procedure specification preparation is therefore the completion of the test design specification, system/software requirement specification, software architecture specification, interface control document and software design description.

B.4.3. Use

The test procedure specification is a simple set of steps to be followed by the tester and used in conjunction with test case data to verify the expected input/output behaviour of a test item.

Software Test Plans

B.4.4. Format & content

B.4.4.1. Cover page

The cover page shall include:
- Project Title
- Document Number
- Volume Number
- Document Class: `Test Procedure Specification'
- System Name
- Revision Number
- Author
- Approval Authority
- Approval Signature
- Approval Date

B.4.4.2. Page marking

Each page is to be marked with:
- Document Class: `Test Procedure Specification'
- System Name
- Document Number
- Revision Number
- Page Number

B.4.4.3. *Identification and Purpose (section 1)*

As per the Test Plan document

B.4.4.4. *Purpose (section 2)*

Describe the purpose of the test case in terms of:

- Test items to be tested.
- Expected outcomes.
- Test cases to be exercised.
- References to test item documentation.

B.4.4.5. *Special requirements (section 3)*

Specify any special requirements that must be satisfied for the correct execution of the procedure. For example:

- prerequisite procedures.
- environmental requirements.
- special skills.

B.4.4.6. *Procedure steps (section 4)*

Describe test steps in terms of:

- **Set Up:** the sequence of actions required to prepare for the procedure.
- **Start**: the actions necessary to start the procedure.

Software Test Plans

- **Proceed**: actions necessary during procedure execution.
- **Measurement**: the process of taking test measurements.
- **Shut Down:** the actions necessary to suspend testing when unscheduled events dictate.
- **Restart**: restart points and the actions necessary to restart the procedure at these points including the management of intermediate data sets produced.
- **Stop**: the actions necessary to bring test execution to an orderly halt.
- **Wrap Up**: the actions necessary to restore the environment.
- **Contingencies**: the actions necessary to deal with anomalous events which may occur during execution.
- **Logs**: the process of logging test results (refer sections 7, 8 and 9).

B.5. Test Item Transmittal report

B.5.1. Purpose

To specify the test items being transmitted for testing.

B.5.2. Software life cycle context

This report documents the handover of the test items from the developer to the tester.

B.5.3. Use

This report basically acts as a check to confirm that the test item is ready for testing.

B.5.4. Format & content

B.5.4.1. Cover page

The cover page shall include:
- Project Title
- Document Number

Software Test Plans

- Volume Number
- Document Class: `Test Item Transmittal Report'
- System Name
- Revision Number
- Author
- Approval Authority
- Approval Signature
- Approval Date

B.5.4.2. Page marking

Each page is to be marked with:

- Document Class: `Test Item Transmittal Report'
- System Name
- Document Number
- Revision Number
- Page Number

B.5.4.3. Identification (section 1)

As per the Test Plan document

B.5.4.4. Transmitted Items (section 2)

Describe test items being transmitted for testing, include the following:

- Version/revision number.
- References to item documentation.
- The related test plan.
- People responsible for the test item.

B.5.4.5. Location (section 3)

Identify the location of the test items to be transmitted. Include the following:

- The media that contains the items
- How the media is labeled or identified

B.5.4.6. Status (section 4)

Describe the status of the test items. Include the following:

- Describe any deviations from the test items documentation.
- Describe any changes from the previous transmittal of the test item.
- List the incident report(s) that are expected to be resolved with this transmittal.
- Any pending modification that may affect the test item

B.5.4.7. Approvals (section 5)

Identify all the names and position of the people who are to approve the transmittal and reserve a space for signatures.

B.6. Test log

B.6.1. Purpose

To provide a chronological record of events that occur in the execution of a test.

B.6.2. Software life cycle context

The Test Log is created during test execution.

B.6.3. Use

The Test Log is used:

- by the tester to record the results of testing and the hardware/software context in which the test was run
- by the Quality Control function to verify that the test was run and was a faithful implementation of approved Test Plans and Test Design Specifications
- by the test manager to ensure that a test can be repeated
- by the maintenance engineer as an aid to rectification of faults.

Software Test Plans

B.6.4. Format & content

B.6.4.1. Cover page

The cover page shall include:

- Project Title
- Document Number
- Volume Number
- Document Class: `Test Log'
- System Name
- Revision Number
- Author
- Approval Authority
- Approval Signature
- Approval Date

B.6.4.2. Page marking

Each page is to be marked with:

- Document Class: `Test Log'
- System Name
- Document Number
- Revision Number
- Page Number

Software Test Plans

B.6.4.3. Identification (section 1)

As per the Test Plan document.

B.6.4.4. Description (section 2)

Provide information that applies to all log entries. For example:

- test item identification including item number and revision level
- details of the actual test hardware/software environment (e.g. memory size, processor type and serial number, peripherals and mass storage devices, test rigs and system software).

B.6.4.5. Activity & event entries (section 3)

Record the following mandatory information for each event:

- date and time
- author
- test procedure identifier and staff present.

> Record the following information as applicable:

- **Procedure Results:** visually observable results, for example:
- error messages generated
- failures

Software Test Plans

- requests for operator action
- test pass or fail.
- **Environmental Information:** any test environmental issue specific to this entry (e.g. hardware substitution).
- **Anomalous Events:** record what happened before and after an unexpected event.
- **Test Incident Report Identifiers**: record the identity of all test incident reports generated.

B.7. Test incident report

B.7.1. Purpose

To document any event that occurs during the testing process that requires further investigation.

B.7.2. Software life cycle context

The Test Incident Report is created during test execution.

B.7.3. Use

The Test Incident Report is used:
- by the tester to record faults identified during testing
- by the Quality Control function to gauge the quality and potential reliability and maintainability of the test item
- by the maintenance manager to initiate corrective action
- by the maintenance engineer as an aid to rectification of faults.

Software Test Plans

B.7.4. Format & content

B.7.4.1. Cover page

The cover page shall include:
- Project Title
- Document Number
- Volume Number
- Document Class: `Test Incident Report'
- System Name
- Revision Number
- Author
- Approval Authority
- Approval Signature
- Approval Date

B.7.4.2. Page marking

Each page is to be marked with:
- Document Class: `Test Incident Report'
- System Name
- Document Number
- Revision Number
- Page Number

Software Test Plans

B.7.4.3. Identification (section 1)

As per the Test Plan document

B.7.4.4. Summary (section 2)

Provide:

- a summary of the incident
- the test item involved (including version and revision level)
- the test procedure and test case identifier
- a reference to test log.

B.7.4.5. Incident description (section 3)

Provide a description of the incident in terms of:
- inputs
- expected results
- actual results
- anomalies
- date and time
- procedure step
- environment
- attempts to repeat
- testers and observers present.

B.7.4.6. Impact (section 4)

If known, indicate what impact this incident will have on test plans, test design specifications, test procedure specifications or test case specifications.

B.8. Test Summary report

B.8.1. Purpose

To provide a management summary of test results and to provide evaluations of the quality of test items.

B.8.2. Software life cycle context

The test summary report is prepared on completion of a predefined set of test procedures. The number and frequency of reports is specified in the test plan.

B.8.3. Use

The test summary report is used by:

- development management to monitor the progress of testing and the degree of conformance to customer requirements.
- quality management to monitor product quality, testing effectiveness and general quality management system effectiveness.

B.8.4. Format & content

B.8.4.1. Cover page

The cover page shall include:

- Project Title
- Document Number
- Volume Number
- Document Class: `Test Summary Report'
- System Name
- Revision Number
- Author
- Approval Authority
- Approval Signature
- Approval Date

B.8.4.2. Page marking

Each page is to be marked with:

- Document Class: `Test Summary Report'
- System Name
- Document Number
- Revision Number
- Page Number

B.8.4.3. Identification (section 1)

As per the Test Plan document

B.8.4.4. Summary (section 2)

For each test item provide:

- the test item identifier including version/revision level
- a summary test item evaluation
- the test environment
- references to test plans, test design specifications, test procedure specifications and test case specifications, test logs and test incident reports.

B.8.4.5. Variances (section 3)

Report any variances of test items from design specifications.

Report any variances from test plans, designs, procedures or cases and provide reasons for each variance.

B.8.4.6. Comprehensive assessment (section 4)

Evaluate the comprehensiveness of the test process against the criteria specified in the test plan. Identify features that were not sufficiently tested and provide reasons.

B.8.4.7. *Summary of results (section 5)*

Summarise the test results in terms of:

- test incidents that have been resolved - summarising the resolution
- test incidents that have not been resolved.

B.8.4.8. *Evaluation (section 6)*

Provide an overall evaluation of each item in terms of:

- limitations
- test performance (pass/fail).
- potential reliability and maintainability.

B.8.4.9. *Summary of activities (section 7)*

Summarise major testing activities and events.

Provide test function performance data in terms of:

- resource consumption (e.g. people, machine time, total elapsed time for testing activities)
- performance against plan (e.g. schedule variance, cost variance).

B.8.4.10. Approvals (section 8)

Identify all the names and position of the people who are to approve the test summary report and reserve a space for signatures.

∎

Printed in Great Britain
by Amazon.co.uk, Ltd.,
Marston Gate.